Organic Blood

The Role of Spices and Sulphur as Blood Substitutes in Ancient Greek Sacrificial Rituals

A Peer Reviewed Article

Marija Elektra Rodriguez

HUNTRESS INK

ISBN: 0-9876009-1-5
ISBN-13: 978-0-9876009-1-2

DEDICATION

This book is dedicated to Elurra.
Much of this research was completed while you
were a hope in our hearts and a tickle in my belly.

IV

ACKNOWLEDGEMENTS

This article was originally published in the peer reviewed journal CLASSICVM Vol. XXXVIII.1, 2012.

Thank you to my father, Giovanni Pennisi, for providing the inspiration for this piece with his stories from the Old Country.

Synopsis

This argument attempts to provide a new perspective on the use of spices and minerals in Greek rituals. Specifically, their symbolic association with blood, immortality and cleansing will be evaluated through the review and analysis of primary source materials. The relationship between the gods and the sacred smoke generated by spices and minerals will be explored, as will the structuralist and functionalist theories with relation to sacrifice. The perpetuation of elaborate "hunting" narratives associated with the procurement of spices will also be evaluated in order to gauge Greek attitudes towards these sacred substances. Ultimately, it is evident that the physical resemblance and symbolic association of "liminal" spices and sulphur with sacrificial blood added to the evolution of Greek rituals, assisting in the connection of worshippers with their gods.

Organic Blood

The Role of Spices and Sulphur as Blood Substitutes in Ancient Greek Sacrificial Rituals

The burning of spices and minerals played a key role in ancient Greek rituals associated with cleansing and sacrifice.[1] These substances represented a cross-cultural connection to ceremonial practice

---

[1] Spices, by definition, refer to seeds, fruits or the woody barks of plants. See Oztekin, Z. & Martinov, M. *Medicinal and Aromatic Crops: Harvesting, Drying and Processing* (New York: CRC Press, 2007), p.2. Minerals, such as sulphur and sea salt, are substances which have a crystalline atomic structure and specific chemical composition. Minerals were also used in Greek cooking, rituals and ointments in antiquity. See G. Rapp, *Natural Science in Archaeology: Archaeo-mineralogy* (Duluth: Springer-Verlag, 2009) pp.17-19.

as the majority of spices used in ritual were not native to Greece, but, rather, were associated with the geographically distant lands of Arabia, the Levant, and Asia.[2] Those which were domestically obtainable, such as mastic and the mineral sulphur, were found in isolated areas and in small quantities.[3] This raises the

---

[2] C. Corn, *The Scents of Eden: A History of the Spice Trade* (New York: Kodansha America Inc., 1999) p.202. Detienne discusses the Greek belief, as conveyed by Theophrastus, that many spices were native to the Arabian Peninsula. In actuality, cinnamon and cassia were native to China and Indonesia. See M. Detienne, *Gardens of Adonis*, trans. Janet Lloyd (Princeton: Princeton University Press, 1994) p.9, 16; Theophrastus, *Enquiry into Plants II*, trans. Sir Arthur Hort (London: William Heinemann Ltd, 1916) Book 9.4.1-2, pp.233-5.

[3] Mastic is native to the island of Chios and was harvested in ancient times, and sulphur occurs naturally in volcanic sites. For Mastic see C.L. Mantell, "The Natural Hard Resins: Their Botany, Sources and Utilization"

question as to why the Greeks used such substances in their rituals and what significance they attached to burning spices and minerals. Were these spices purely functional, or did the Greeks ascribe a deeper, representational meaning with their use? Primary source material, spanning from the Archaic to Hellenistic

---

Economic Botany 4 no.3 (Jul –Sep, 1950), p.240. Theophrastus gives a general survey of plants which produce resin around the Peloponnesus and Crete and discusses attempts to harvest myrrh. See Theophrastus, *Enquiry into Plants II,* 9.1.2-5, p.219-20. For volcanic sulphur in the Mediterranean see G.F. Rodwell, *Etna: A History of the Mountain and its Eruptions* (New York: Cambridge University Press, 2011) p.9, 42, 56. Leighton suggests that archaeological evidence points to sulphur, sourced from Sicily, being traded around the Mediterranean as early as 16th or 17th centuries BC. See R. Leighton, *Sicily before History: An Archaeological Survey from the Palaeolithic to the Iron Age* (Ithaca: Cornell University Press, 1999) p.4.

periods, suggests that the Greeks harboured a strong, symbolic association of certain spices and minerals with purity, cleansing and communication with their immortal gods. By analysing Detienne's theory that the Greeks viewed spices as substances of the divine realm, and by evaluating the physical properties of these spices and minerals as they burn, it is evident that there was a visual and symbolic connection between sacrificial blood and these substances. In the Greek mind, spices and minerals, through their connection with blood, could act as either a purifying agent or a conduit to the gods.

The burning of spices in sacrificial rituals represented the transition of a vegetative substance into a powerful religious symbol, with strong connections to both immortality and blood. Detienne suggests that spices were the foods of the immortals as they were cooked by flames, but left no residual for consumption by mortals.[4] Rather, they transitioned to the gods via smoke, unlike animal flesh, which remained behind for the worshippers.[5] Spices came from the eastern lands, which, in Classical Greek thought, were considered to be favoured by the sun.[6] Detienne associates a dry, burning creation process with immortality, the antithesis of "wet" substances, which were associated with death and decay.[7] However,

---

[4] Detienne, *Gardens of Adonis,* p.45, 49.

[5] The smell of the meat was also considered to rise to the gods. Detienne, *Gardens of Adonis,* p.45, 49.

[6] Detienne, *Gardens of Adonis,* pp.9-10.

[7] Detienne, *Gardens of Adonis,* pp.9-10, referring to Herodotus' description. See Herodotus, *Histories,* trans. Aubrey de

Detienne treats all burned spices in the same fashion, and does not distinguish between those that were burned dryly, such as cassia and cinnamon, and those that transformed into a liquid as they generated smoke.[8] This subcategory of spices is liminal in nature; they acted as transitional substances that exhibited a symbolic resemblance to sacrificial blood, the very substance that Burkert argues

---

Sélincourt (London: Penguin Books, 1972) Book 3.113, p.249-50.

[8] The other substances that Detienne uses in his examples, such as cinnamon, cassia, and the herbs laurel, thyme and sage, all burn "dryly" and do not create a liquid. See Detienne, *Gardens of Adonis*, pp.38-9. For the melting of myrrh and resin when heated see Harlan, R., & Gannal, J.N. *History of Embalming and of Preparations in Anatomy, Pathology and Natural History* (Charleston, BiblioLife LLC, 2009) p.100. Theophrastus also discusses myrrh existing in liquid form. See Theophrastus, *Enquiry into Plants II*, 9.4.10, p.241.

emphasises the sanctity of human life during sacrifice.[9] In particular, frankincense, myrrh, and mastic liquefied when burned, suggesting a form of ritualistic, "organic" blood.

The symbolic association of myrrh and blood began with the harvesting of the myrrh tree; a highly ceremonial ritual alluded to in the Adonis myth and echoed

---

[9] W. Burkert, Homo *Necans: The Anthropology of Ancient Greek Sacrificial Ritual and Myth,* trans. Peter Bing (Berkeley, Los Angeles & London: University of California Press, 1983) p.38. McCarthy provides an alternative interpretation on the symbolism of blood in Greek sacrificial ritual, arguing that it did not belong to the gods, but rather, was primarily associated with "cults of the dead" and animating *eidolons.* McCarthy's argument neglects to evaluate the belief in the importance of smoke in communicating with the gods during sacrifice. See D.J. McCarthy, "The Symbolism of Blood and Sacrifice" *Journal of Biblical Literature* 88 no.2 (June, 1969) p.254.

in blood-sacrifice rituals. Tree sap, the fluid from which myrrh is made, is analogous to blood: it distributes nutrients throughout the tree, just as blood circulates nutrients throughout the body.[10] This fact may not have been widely understood by Classical Greeks, however, at its simplest level, the cutting of a tree to spill the sap for collection is visually analogous to the cutting of human flesh to spill blood.[11] The husbandry rites of myrrh were highly ritualistic, and were

---

[10] For sap as circulating nutrients see T. Wessels, *Forest Forensics: A Field Guide to Reading the Forested Landscape* (Woodstock: The Countryman Press, 2010) p.157.

[11] Theophrastus alludes to nutrients as he discusses the importance of sap and resin to plant growth, and also details which saps are fattier in nature than others. See Theophrastus, *Enquiry into Plants II*, 9.1.1-7, pp.217-21.

passed on to exclusive, "sacred" families.[12]
The sap was harvested from chosen trees
in sacred groves, and one third of the
substance was dedicated to the Sabaean
sun god.[13]   Those who obtained the
sacred substance were required to be pure;
intercourse and contact with the dead
were forbidden prior to the procurement
ritual.[14]   During the extraction process,
the swollen myrrh tree was struck with an
axe in a ceremonial fashion, bearing a
resemblance to the Adonis birth myth in
which Smyrna/Myrrha, being transformed
into   a   tree,   was   split   open   during

---

[12] The procurement of myrrh was known to
the Greeks and is described by Theophrastus
writing from the late Classical period.   See
Theophrastus, Enquiry *into Plants II,* 9.4.1-10,
pp.233-41; Detienne, *Gardens of Adonis,* pp.6-7.
[13] Theophrastus, *Enquiry into Plants II,* 9.4.4-6,
pp.237-9; Detienne, *Gardens of Adonis,* pp.6-7.
[14] Detienne, *Gardens of Adonis,* pp.6-7.

childbirth; and, also, the Greek sacrificial ritual in which the animal, especially selected and prepared for death, was killed by an axe.[15] Initially, the harvested sap was clear; however, as the substance became depleted, it turned into a blood-red liquid.[16] The small, round beads of resin which seep from the myrrh tree were

---

[15] Apollodorus, writing from the Hellenistic period, recounts Panyassis' account of the myth from the Archaic period. See Apollodorus, *The Library of Greek Mythology*, trans. Robin Hard (New York & Oxford: Oxford University Press, 1998) Book 3.14.4, p.131. For the axe wounds inflicted on myrrh trees during harvesting, see Theophrastus, *Enquiry into Plants II*, 9.4.4, pp.237. For the process of Greek blood-sacrifice and the ceremonial striking of the animal with an axe see Homer, *The Odyssey*, trans. Walter Shewring (New York & Oxford: Oxford University Press, 2008) Book 3.447-52, p.33.

[16] Detienne, *Gardens of Adonis*, p.8. Theophrastus also discusses blood-red plant resin. See Theophrastus, *Enquiry into Plants II*, 9.1.1, p.217.

likened to Myrrha's tears by Ovid, writing from a Roman perspective in the first century AD; however, there is no literary evidence to suggest that the earlier Greeks viewed it as such.[17] Schoff recounts an Arabian association of the drops of tree gum with clots of menstrual blood, highlighting the association of trees with the body and sap with blood by those who harvested the spice.[18] Such an association may have been passed on to the Greeks with the introduction of the substance into Greek ritual.[19] Myrrh was considered

---

[17] Ovid, *Metamorphoses*, trans. David Raeburn (London: Penguin Books, 2004) Book 10.499-501, p.407.

[18] W.H. Schoff, (trans.), *The Periplus of the Erythraean Sea: Travel and Trade in the Indian Ocean by a Merchant of the First Century* (New York: Longmans, Green and Co., 1912) p.131.

[19] Burkert argues that the Adonis cult was connected to the Semitic world by the import

particularly sacred to Aphrodite and was burned on her altars; it provided a vegetative parallel to sacrificial animal blood as the two substances comingled or burned alongside one another.[20] This further suggests a strong symbolic connection existed between blood and myrrh, which, when burned, reverted into a liquid substance.

---

of myrrh, around the seventh century BC. See W. Burkert, *Structure and History in Greek Mythology and Ritual* (Berkeley & Los Angeles: University of California Press Ltd, 1979) p.106.

[20] For the importance of sweet smell and incense being sacred to Aphrodite see M.S. Stoddart, *The Scented Ape: The Biology and Culture of Human Odour* (Cambridge: University of Cambridge Press, 1990) p.177. For *haimassein,* the sacredness of spilling animal blood on altars, see W. Burkert, *Greek Religion: Archaic and Classical,* trans. John Raffan (Malden, Oxford & Carlton: Blackwell Publishing Ltd, 1985) p.56.

Ancient Greek beliefs associated "hunting" narratives with spice harvesting, providing a parallel between the shedding of animal blood and the shedding of tree resin. Burkert, from a functionalist perspective, argues that sacrifice is primarily an act of killing with its roots in hunter-gather society.[21] His interpretation of sacrifice highlights the Greek belief that ritualised animal slaying was one of many pious acts through which humankind could appease, and "experience" the gods.[22] It is the act of killing an animal, according to Burkert, that signifies the sacredness of life; moreover, the elaborate ritual stages helped to mitigate the guilt associated with bloodshed.[23] By

---

[21] Burkert, Homo *Necans,* p.3, 14.

[22] Burkert, *Homo Necans,* p.2, 40.

[23] Burkert, *Homo Necans,* p.38.

contextualizing the sacrificial act in this manner, the burning of spices deepens the connection to the "hunting" roots of the ritual, as many of the key spices had stories of elaborate procurement methods associated with their harvesting.[24] Herodotus discusses the complexity of obtaining frankincense by the Arabians, who had to burn storax, a form of gum, in order to ward away the "flying snakes" that protected the sacred substance.[25] As

---

[24] Herodotus, *Histories,* 3.107-12, pp.248-9.

[25] Herodotus, *Histories,* 3.107, p.248. Herodotus also gives the example of obtaining cassia, which he recounts as being protected by bat-like creatures which assaulted the spice collectors. Cinnamon, which grew "somewhere in the region where Dionysus was brought up", was stolen from the nests of giant birds by distracting the creatures with oxen or donkey meat. Ladanon, a form of mastic, was believed to have been gathered from the beards of he-goats. See Herodotus, *Histories,* 3.111-12, p.249. Herodotus states

Schoff argues, Herodotus was possibly recounting an Arabian myth which has its roots in animistic worship, a belief in tree spirits that resembled winged serpents, who possessed and guarded over the sacred vegetation.[26] Such spirits were thought to stalk the merchants of frankincense, as these spirits believed the sap to be their blood, and it was only through the act of burning another

---

that Ladanon is not native to Greece; however, there is evidence to suggest that mastic is native to the island of Chios. See Mantell, "The Natural Hard Resins", p.240.

[26] Schoff, *The Periplus of the Erythraean Sea*, p.130-1. Hutchinson also discusses the Arabian belief in frankincense spirits. See R.W. Hutchinson, "The Flying Snakes of Arabia" *The Classical Quarterly* 8 no.1/2 (May, 1958) pp.100-1. Myrrh, Schoff argues, does not have a "hunting" narrative as it was cultivated in a Joktanite area in which animism did not flourish. See Schoff, *The Periplus of the Erythraean Sea*, p.132.

spiritual "blood", storax, that such creatures could be repelled.[27] The Arabian association of serpents guarding sacred spices echoes the Greek myth of Ladon protecting the apples of the Hesperides, another substance connected to immortality.[28] The extraordinary nature of the collection of frankincense struck some critics, such as Pliny the Elder writing in the first century AD, as a fictional story told by the Arabians in order to inflate the price of their spices.[29] As Detienne illustrates, such readings of

---

[27] Schoff, *The Periplus of the Erythraean Sea,* p.130-2.

[28] Apollodorus, *Library,* 2.5.11, p.81.

[29] Pliny, *Natural History IV,* trans. H. Rackham (London: William Heinemann Ltd, 1952) Book 12.85, p.63. Detienne, *Gardens of Adonis,* p.17. Schoff describes these passages as being "laughed at as travellers' yarns". Schoff, *The Periplus of the Erythraean Sea,* p.131.

Herodotus are problematic; the fantastical stories told about spice harvesting perpetuated the sacredness of the substance and cannot be dismissed as purely economically motivated.[30] The "hunting" narratives associated with these spices parallel the hunting of animals, which Burkert argues is at the origin of the sacrificial ritual.[31] Herodotus' association of spice gathering with wild, elusive, or fantastical creatures emphasises the Greek perception of otherworldliness associated with these sacred substances. The inclusion of spices in the sacrificial practice illustrates the evolution of the sacred rite; spices provided an additional level of complexity to the ceremony, and

---

[30] Detienne, *Gardens of Adonis,* p.17.
[31] Burkert, *Homo Necans,* p.3, 14.

yet still maintained a connection to the "hunting" roots of the ritual.

Sacrifice is argued to be a key sacred ritual that distinguished the Classical Greeks from animals and solidified their place with respect to the gods.[32] The spice burning in this ritual added an additional dimension of complexity to the ceremony, and offers an insight into Greek thought regarding the divine. In Eleusinian myth, the act of burning is strongly connected to immortality and the gods.[33] Demeter ritualistically burned Demophoön, anointed in "sweet smelling" ambrosia, in

---

[32] Detienne, *Gardens of Adonis,* p.xi.
[33] A.N. Anthanassakis (trans.), *Homeric Hymn to Demeter* (Baltimore: The Johns Hopkins University Press, 2004) lines 234-42, p.7.

order to render him immortal.[34] This echoes a similar motif in *Iliad,* in which Thetis anoints Patroclus with ambrosia in order to prevent decay prior to his incineration on the funeral pyre.[35] Apollodorus, writing sometime in the first or second century AD but capturing myths of earlier periods, also gives a description of Thetis anointing Achilles with ambrosia and placing him in flames in order to render him immortal.[36] The use of ambrosia as an ointment suggests that the food of the gods was a spice-like substance, as frankincense and myrrh were key ingredients in many Greek ointments,

[34] Anthanassakis, *Homeric Hymn to Demeter,* 234-42, p.7.

[35] Homer, *The Iliad,* trans. Robert Fitzgerald (New York & Oxford: Oxford University Press, 2008) Book 19.31-48, pp.338-9.

[36] Apollodorus, *Library,* 3.13.6, p.129.

transforming regular olive-oil into a medicinal balm.[37] It is this spice-balm that provided a catalyst for immortality in Greek thought; without the ritualistic anointment, there was no physical substance to connect the fire to their immortal gods. Just as smoke was not perceived to be sacred unless the substance that created it was worthy of the gods, the goddesses were unable to transfer immortality without the incorporation of a "divine" substance. In sacrificial ritual, the inclusion of spices provided the Greeks with a tangible

---

[37] Detienne, *Gardens of Adonis*, p.37, 48, 61. For spices in ointment see Detienne, *Gardens of Adonis,* p.37, 60. For the various medicinal properties of myrrh, including as an antiseptic, see G. Majno, *The Healing Hand: Man and Wound in the Ancient World* (Cambridge: Harvard University Press, 1991) p. 20, 64, 218.

connection to their gods. The burning of substances associated with immortality and sweet smells were perceived to attract the gods and provided a relationship between the divinities and their worshippers.[38] This medium of connection was required due to a segregated hierarchy: gods were the superior beings, followed by the "earth-dwelling" categories of humans and beasts.[39] Animals were seen to eat each

---

[38] Detienne, *Gardens of Adonis,* p.38, with reference to commentary on sacrifice in Aeschines. See Aeschines, *Against Timarchus,* trans. Nick Fisher (New York & Oxford: Oxford University Press, 2001) sect.23, p.76.
[39] J.P. Vernant, *Myth and Thought among the Greeks,* trans. Janet Lloyd & Jeff Fort (Brooklyn: Urzone Inc. 2006) p.56; Detienne, *Gardens of Adonis,* p.xii. Hesiod emphasises the separation of gods and humans in the iron age of man. See Hesiod, "Works and Days", trans. Dorothea Wender, in Dorothea Wender ed. *Hesiod: Theogony and Works and Days &*

other without ceremony; however, the consumption of animal flesh by humans was differentiated as it was enveloped by ritual in honour of the gods.[40] The structuralist perspective argues that rituals associated with communal eating were precisely what distinguished humans from beasts.[41] The inclusion of spices into this hierarchy highlights this distinction further: Greek worshippers incorporated an additional paradigm, that of vegetation, through the blending of organic, "blood-like" gums, or dry, "immortal" spices, with the blood of animals in sacrificial ritual. The burning of spices illustrates the symbolic connection of vegetative

---

Theognis: Elegies (New York: Penguin Books Ltd, 1973) lines 176-202, p.64.

[40] Vernant, *Myth and Thought among the Greeks,* p.99.

[41] Detienne, *Gardens of Adonis,* p.xi.

substances to immortal gods in sacrifice, just as the anointment and burnings of human flesh with spices, in Greek myth, could potentially render a human immortal.

In pre-classical literature, the burning of mineral sulphur was associated with purification, cleansing rituals and the ability to remove the pollution associated with *miasma*. Homer's *Odyssey* reflects an Archaic period perception that sulphur was considered an appropriate cleanser for murder.[42] After slaughtering the suitors, Odysseus requested sulphur in order to "cleanse away this pollution", referring to *miasma*, the contamination associated with

---

[42] Homer, *Odyssey*, 22.479-95, p.276.

murder.[43] He then proceeded to burn sulphur, reducing the yellow coloured mineral to a blood-red liquid, in order to cleanse his palace.[44] Eurycleia acknowledges that Odysseus' request is fitting for the situation, she states "all that you say is well said", validating the appropriateness of sulphur as a cleansing agent.[45] In this instance, sulphur, a form of "organic" blood, acts as a substitute for pig's blood, the purifying substance used to mitigate *miasma*.[46] In such rituals, the

---

[43] Homer, *Odyssey*, 22.480-3, p.276; R. Parker, *Miasma* (Oxford: Oxford University Press, 2001) p.39.

[44] Homer, *Odyssey*, 22.290-5, p.276. For the chemical transition of sulphur into a red, viscous liquid when melted see R.J.W. Cremlyn, *An Introduction to Organosulfur Chemistry* (Chichester: John Wiley & Sons Ltd, 1996) p.7-8.

[45] Homer, *The Odyssey*, 22.486-90, p.276.

[46] Parker, *Miasma,* pp.370-3.

blood of a pig was used to cleanse a suppliant of murder, as in the case of Orestes or Jason.[47]   As Parker suggests, the sacrificial pig became the object of the vendetta, and thereby mitigated the need for further bloodshed.[48]   Apollodorus illustrates this in his description of Jason's ritualistic cleansing by Circe.[49]   In this instance, Circe is able to propitiate Zeus' anger towards Jason for the murder of Apsyrtus by washing his hands with the

---

[47] Parker, *Miasma,* pp.371-2. For the depiction of Orestes being cleansed by pig's blood on Attic vases see Ziolkowski, T. *The Mirror of Justice: Literary Reflections of Legal Crises* (Princeton: Princeton University Press, 2003) pp.33-4, 278.

[48] Parker, *Miasma,* pp.372.

[49] Apollonius, *Jason and the Golden Fleece,* trans. Richard Hunter (New York & Oxford: Oxford University Press, 2009) lines 4.702-19, p.115.

blood of a newly born pig.[50] This illustrates the way in which Greek culture perceived that sulphur, when burned in purification rituals and converted into a blood-like substance, provided a connection to the divine notion of purity and was an acceptable substitute for pig's blood in removing *miasma*.

The association of sulphur as a purifying agent is further conveyed in the Classical period, as the mineral substance was used in general cleansing rituals.[51] In Euripides' *Helen*, Theonoe orders her

---

[50] Apollonius, *Jason and the Golden Fleece,* 4.699-715, p.115.

[51] The Judeo-Christian association of sulphur with "evil" is in vibrant contrast to the Greek connotation of sulphur with cleansing and purity. See A.E. Bernstein, *Formation of Hell: Death and Retribution in the Ancient and Early Christian Worlds* (New York: Cornell University Press, 1993) p.259.

servants to purify the palace with burning sulphur before she enters, so that she might "breathe heaven's pure air".[52] As she suggests, the Greeks perceived the smoke created by burning sulphur as purifying, transforming the potentially tainted air of the palace into a sacred atmosphere.[53] In its simplest form, the

---

[52] Euripides, "Helen", trans. David Kovacs, in David Kovacs ed. *Loeb Classical Library: Euripides Helen, Phoenician Women, Orestes* (London: Harvard University Press, 2002) lines 865-72, pp.112-5.

[53] Allan highlights the redundancy of some scholarly claims, such as that of Mikalson, that Theonoe's instruction to burn sulphur was an Egyptian custom, and, by implication, not Greek. Although many aspects of Theonoe's mannerisms are implied to be Egyptian, the practice of burning sulphur was also a Greek custom. See Allan's commentary on Euripides, *Helen,* ed. William Allan (Cambridge: Cambridge University Press, 2008) p.243, with reference to lines 865-72. J.D. Mikalson, *Honour Thy Gods: Popular Religion in Greek Tragedy* (Chapel Hill: The

potent smell of sulphur smoke could mask unfavourable odours and prepare a space for sacred ritual.[54] Greeks believed that strong smells had the ability to ward off impurities connected with death, such as

---

University of North Carolina Press, 1991) p.97.

[54] The smell of sulphur compounds varies greatly. Some forms of gaseous sulphur, such as hydrogen sulphide or calcium sulphide, smell like "rotten eggs". See T.W.G. Solomons *Organic Chemistry* (Chichester: John Wiley & Sons Ltd, 2011) p.881. However, when sulphur crystals or powder are burned they form sulphur dioxide, which has a potent, but not rotten, smell. See H. Perkins-Cady, *General Chemistry* (Columbus: McGraw-Hill Books Co. Inc., 2010) p.5, 171. The Homeric and Archaic Greeks most likely used sulphur in its crystalline or powdered forms, which occurred naturally near volcanic sites, such as Mt. Etna in the colony of Sicily. Rodwell, *Etna: A History of the Mountain and its Eruptions,* p.9, 56. Homer mentions the use of powdered crystalline sulphur in religious cleansing. See Homer, *Iliad,* tr. Fagles, 16.270, p.420.

infestation by insects or malevolent spirits associated with *miasma*.[55] However, there were many strong smelling substances which could have functioned just as well, if not better, than sulphur. For instance, mastic, frankincense and myrrh each produced sweet smelling smoke and were also desirable in ritual if the objective was merely to sweeten, or alter, the scent of the air.[56] Parker suggests that fumigant sulphur acted as an airborne disinfectant; however, the actual intensity of this cleanser was so mild that, from a scientific perspective, it was more symbolic than functional.[57] An alternative explanation

---

[55] *Miasma* was associated with attracting hostile, vengeful spirits of murder victims. Parker, *Miasma*, p.107. For strong smells as deterrents see Parker, *Miasma*, p.231.
[56] Detienne, *Gardens of Adonis*, p.26.
[57] Parker, *Miasma*, pp.58, 227-8.

for the desirability of sulphur in cleansing rituals is its physical similarity to blood when it is introduced to flames.[58] Liquid sulphur not only resembled blood in its colour, but also in its viscosity, giving it the appearance of recently oxygenated or coagulated blood, intensifying the resemblance to the ceremonial animal blood used in *miasma* cleansing.[59]

Sacrificial blood was also used alongside burning sulphur for purification, as illustrated in Euripides' *Iphigenia among*

---

[58] Cremlyn, *An Introduction to Organosulfur Chemistry*, p.7-8.

[59] For the increase in viscosity of sulphur compounds when heated see A. Earnshaw & Greenwood, N. *Chemistry of the Element, Second Edition* (Oxford: Elseveir Science Ltd, 2002) p.683. For the thickening and coagulation of blood once it is exposed to air see C. Winner, *Circulating Life: Blood Transfusion from Ancient Superstition to Modern Medicines* (Minneapolis: Twenty-First Century Books, 2007) p.24.

the Taurians.[60]   Iphigenia states that she
would use the blood of newborn lambs
and torches to "wash away the tainting
bloodshed" and cleanse the goddess
Artemis of the pollution associated with
murder and death.[61] The incorporation of
both sacrificial blood and burning sulphur
suggests a symbolic relationship which was
also paralleled in the physical resemblance
of these two substances. Detienne
suggests that sacred smoke provided a
communication "channel" to the gods,

---

[60] Euripides, "Iphigenia among the Taurians",
trans. David Kovacs, in David Kovacs ed.
*Loeb Classical Library: Euripides Trojan Women,
Iphigenia among the Taurians, Ion* (London:
Harvard University Press, 1999) lines 1210-34,
pp.283-9.
[61] Euripides, *Iphigenia among the Taurians*, 1223-
9, p.287. See also Allan's commentary on
sulphur in Euripides, *Helen,* p.243 for
validation of the use of sulphur in Iphigenia's
purification ritual.

and the substances which altered the scent of this smoke were worthy of the immortals.[62] Although he does not explore all variations of sacred smoke, he does emphasise that the substances that were burned had to possess a religious significance in order to validate a connection to the divine.[63] This contrasts with Parker's exploration of sacred smoke, as he suggests that sulphur fumes were believed to repel malevolent elements associated with pollution, rather than

---

[62] Detienne, *Gardens of Adonis*, p.37, 61.

[63] Detienne primarily focuses on smoke as connecting worshippers with divinities. See Detienne, *Gardens of Adonis,* pp.38-9. However, he does draw a distinction between the smoke of meat, which emphasises distance between the gods and humankind, and the smoke of spices which bonds them. See Detienne, *Gardens of Adonis,* pp.48-9. For strong smelling substances which attract the Gods see Detienne, *Gardens of Adonis,* p.127.

attract divinities.[64] The use of sulphur and spices suggests that Greek thought encompassed both of these paradigms with relation to sacred smoke: it could represent a communication channel to the gods, or a supernatural repellent - depending on which substance was burned in the ritualistic context.

Rituals provide a community with the ability to bond over common beliefs and shared customs, and often fulfil the emotional, spiritual, or psychological needs of individuals within that society.[65] Religious symbols convey spiritual meaning and signify aspects of the ritual which hold a particular significance within

---

[64] Homer, *Odyssey,* 22.480-3, p.276. Parker, *Miasma,* p.39.

[65] R.L. Grimes, *Beginnings in Ritual Studies* (Oxford: Oxford University Press, 2010) p.i.

the culture. In Greek thought, the act of burning spices and minerals provided a physical reminder of the reverence associated with sacrificial blood, both as a cleansing substance and a conduit to the gods. Sulphur, mastic, myrrh and frankincense each represented key substances that were burned during rituals, and provide an insight into the Greek thought process associated with connection to their gods. In Greek belief, the very act of burning a substance irrevocably changed it; the consumption of blood, fat, bones, and spices by flames transformed stone altars and temples into active centres of connection with the divine. The sweet smell of these substances was believed to attract the gods and assisted in the preparation of worshippers as they entered a sacred space; whereas the potent smoke associated with mineral sulphur repelled creatures connected with contamination. The procurement methods of these various spices were disseminated in myth-like stories that validated these materials as sacred substances, suggesting their ability to symbolically purify religious spaces and cleanse worshippers prior to their rituals. It is the physical, symbolic, mythological and ritual association of these spices and minerals with blood that signifies the sacred role that these substances played in Greek ceremony and thought.

Bibliography

Primary Sources

Aeschines, *Against Timarchus*, trans. Nick Fisher (New York & Oxford: Oxford University Press, 2001).

Anthanassakis, A.N., (trans.), *Homeric Hymn to Demeter* (Baltimore: The Johns Hopkins University Press, 2004).

Apollodorus, *The Library of Greek Mythology*, trans. Robin Hard (New York & Oxford: Oxford University Press, 1998).

Apollonius, *Jason and the Golden Fleece*, trans. Richard Hunter (New York & Oxford: Oxford University Press, 2009).

Euripides, "Helen", trans. David Kovacs, in David Kovacs ed. *Loeb Classical Library: Euripides Helen, Phoenician Women,*

Orestes (London: Harvard University Press, 2002) pp.12-202.

Euripides, *Helen,* ed. William Allan (Cambridge: Cambridge University Press, 2008).

Euripides, "Iphigenia among the Taurians", trans. David Kovacs, in David Kovacs ed. *Loeb Classical Library: Euripides Trojan Women, Iphigenia Among the Taurians, Ion* (London: Harvard University Press, 1999) pp.152-312.

Herodotus, *Histories,* trans. Aubrey de Sélincourt (London: Penguin Books, 1972).

Hesiod, "Works and Days", trans. Dorothea Wender, in Dorothea Wender ed. *Hesiod: Theogony and Works and Days & Theognis: Elegies* (New York: Penguin Books Ltd, 1973).

Homer, *The Iliad*, trans. Robert Fitzgerald (New York & Oxford: Oxford University Press, 2008).

Homer, *The Odyssey*, trans. Walter Shewring (New York & Oxford: Oxford University Press, 2008).

Ovid, *Metamorphoses*, trans. David Raeburn (London: Penguin Books, 2004).

Pliny, *Natural History IV,* trans. H. Rackham (London: William Heinemann Ltd, 1952).

Schoff, W.H. (trans.), *The Periplus of the Erythraean Sea: Travel and Trade in the Indian Ocean by a   Merchant of the First Century* (New York: Longmans, Green and Co., 1912).

Theophrastus, *Enquiry into Plants II,* trans. Sir Arthur Hort (London: William Heinemann Ltd, 1916).

Bibliography

Secondary Sources

Bernstein, A.E. *Formation of Hell: Death and Retribution in the Ancient and Early Christian Worlds* (New York: Cornell University Press, 1993).

Burkert, W. *Greek Religion: Archaic and Classical,* trans. John Raffan (Malden, Oxford & Carlton: Blackwell Publishing Ltd, 1985).

Burkert, W. *Homo Necans: The Anthropology of Ancient Greek Sacrificial Ritual and Myth,* trans. Peter Bing (Berkeley, Los Angeles & London: University of California Press, 1983).

Burkert, W. *Structure and History in Greek Mythology and Ritual* (Berkeley & Los Angeles: University of California Press Ltd, 1979).

Cremlyn, R.J.W. *An Introduction to Organosulfur Chemistry* (Chichester: John Wiley & Sons Ltd, 1996).

Corn, C. *The Scents of Eden: A History of the Spice Trade* (New York: Kodansha America Inc., 1999).

Detienne, M. *Gardens of Adonis*, trans. Janet Lloyd (Princeton: Princeton University Press, 1994).

Earnshaw, A. & Greenwood, N. *Chemistry of the Element, Second Edition* (Oxford: Elseveir Science Ltd, 2002).

Grimes, R.L. *Beginnings in Ritual Studies* (Oxford: Oxford University Press, 2010).

Harlan, R., & Gannal, J.N. *History of Embalming and of Preparations in Anatomy, Pathology and Natural History* (Charleston, BiblioLife LLC, 2009).

Hutchinson, R.W. "The Flying Snakes of Arabia" *The Classical Quarterly* 8 no.1/2 (May, 1958) pp.100-1.

Leighton, R. *Sicily before History: An Archaeological Survey from the Palaeolithic to the Iron Age* (Ithaca: Cornell University Press, 1999).

Majno, G. *The Healing Hand: Man and Wound in the Ancient World* (Cambridge: Harvard University Press, 1991).

Mantell, C.L. "The Natural Hard Resins: Their Botany, Sources and Utilization" *Economic Botany* 4 no.3 (July – September, 1950), pp.203-42.

McCarthy, D.J. "The Symbolism of Blood and Sacrifice" *Journal of Biblical Literature* 88 no.2 (June, 1969), pp.166-76.

Mikalson, J.D. *Honour Thy Gods: Popular Religion in Greek Tragedy* (Chapel Hill: The University of North Carolina Press, 1991).

Oztekin, Z. & Martinov, M. *Medicinal and Aromatic Crops: Harvesting, Drying and Processing* (New York: CRC Press, 2007).

Parker, R. *Miasma* (Oxford: Oxford University Press, 2001).

Perkins-Cady, H. *General Chemistry* (Columbus: McGraw-Hill Books Co. Inc., 2010).

Rapp, G. *Natural Science in Archaeology: Archaeo-mineralogy* (Duluth: Springer-Verlag, 2009).

Rodwell, G.F. *Etna: A History of the Mountain and its Eruptions* (New York: Cambridge University Press, 2011).

Solomons, T.W.G. *Organic Chemistry* (Chichester: John Wiley & Sons Ltd, 2011).

Stoddart, M.S. *The Scented Ape: The Biology and Culture of Human Odour* (Cambridge: University of Cambridge Press, 1990).

Vernant, J.P. *Myth and Thought among the Greeks,* trans. Janet Lloyd & Jeff Fort (Brooklyn: Urzone Inc. 2006).

Winner, C. *Circulating Life: Blood Transfusion from Ancient Superstition to Modern Medicines* (Minneapolis: Twenty-First Century Books, 2007).

Wessels, T. *Forest Forensics: A Field Guide to Reading the Forested Landscape* (Woodstock: The Countryman Press, 2010).

Ziolkowski, T. *The Mirror of Justice: Literary Reflections of Legal Crises* (Princeton: Princeton University Press, 2003).

ABOUT THE AUTHOR

Marija Elektra Rodriguez has a degree in Economics from the University of Queensland and has completed postgraduate studies in this field. She has also completed studies in Literature and Classical Greek History.

In 2011 she was the recipient of the following awards from the University of Sydney:

Walter Reid Memorial Prize (awarded to the top one per cent of students across the Law and Arts faculties).

The Dean's Merit Award (honour roll, awarded to the top scoring students in each faculty).

Classics Essay Prize

Marija is a writer of horror and erotica fiction. Her stories have been widely anthologised in both Australia and internationally.

She lives in Sydney with her husband (El Carnicero), her daughter, and a bunch of pirate pets.